Members of the Animal Kingdom

FISH

TRACY VONDER BRINK

A Crabtree Crown Book

Crabtree Publishing
crabtreebooks.com

School-to-Home Support for Caregivers and Teachers

This appealing book is designed to teach students about core subject areas. Students will build upon what they already know about the subject, and engage in topics that they want to learn more about. Here are a few guiding questions to help readers build their comprehensions skills. Possible answers appear here in red.

Before Reading:

What do I know about fish?

- *I know fish are animals.*
- *I know fish live underwater.*

What do I want to learn about this topic?

- *I want to learn how fish get oxygen underwater.*
- *I want to learn how fish are different from other animals.*

During Reading:

I'm curious to know...

- *I'm curious to know how fish swim.*
- *I'm curious to know how fish develop over time.*

How is this like something I already know?

- *I know that fish swim.*
- *I know that fish hatch from eggs.*

After Reading:

What was the author trying to teach me?

- *The author was trying to teach me how fish are different from other animals.*
- *The author was trying to teach me about what fish eat.*

How did the photographs and captions help me understand more?

- *The photographs helped me understand what different fish look like.*
- *The captions gave me extra information about the body parts of fish.*

TABLE OF CONTENTS

Chapter 1

SORTING IT OUT

People have been curious about living things for thousands of years. One way to study living things is to organize them into groups. Scientists call this **classification**. Living creatures are classified based on the **traits** they have in common.

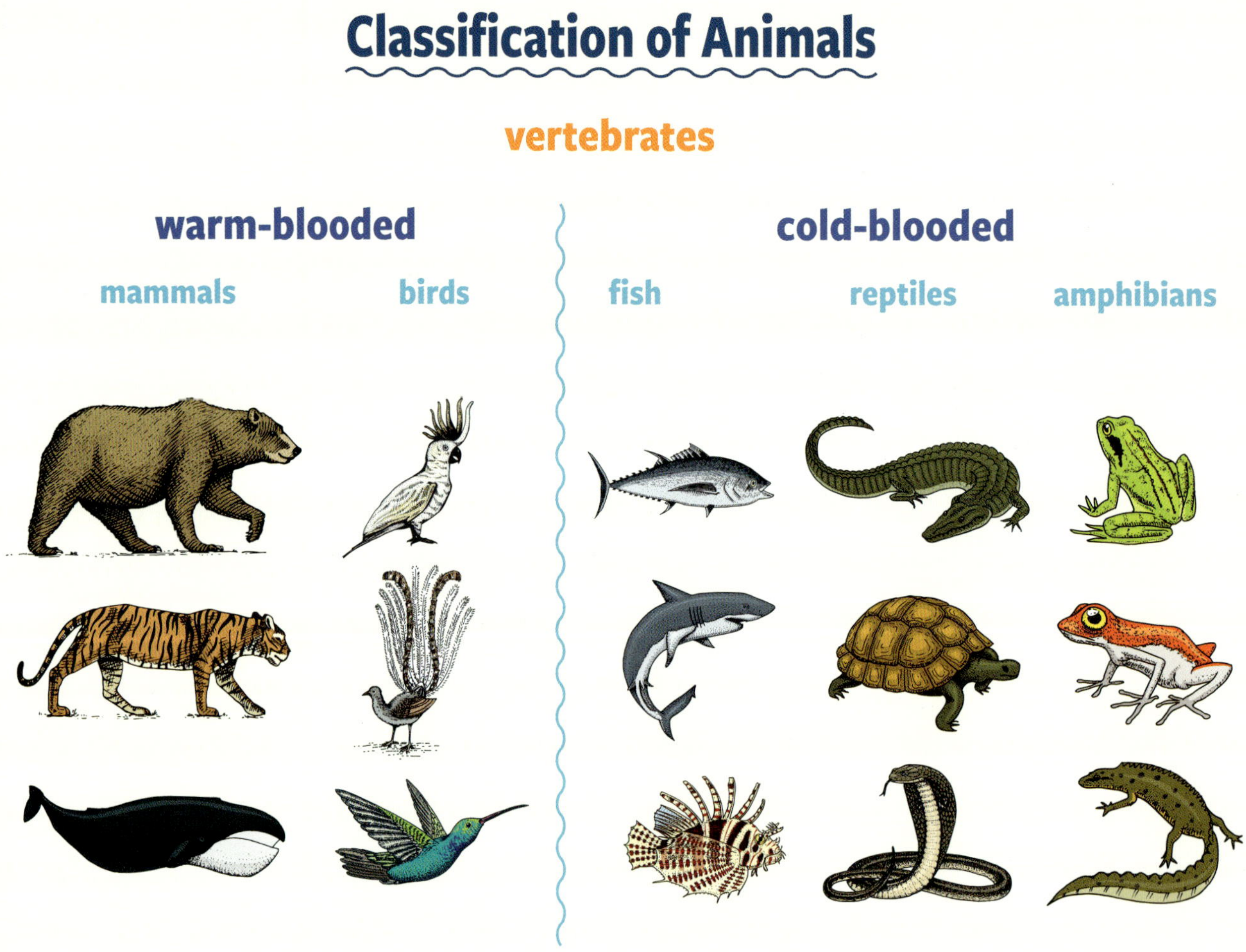

Animals are classified in two main groups. Vertebrates have backbones. Invertebrates do not have backbones.

Living things are divided into kingdoms, such as the animal kingdom. A kingdom contains groups of many living creatures. Fish are one group in the animal kingdom. Each group in a kingdom is then divided into smaller groups. For example, bony fish and jawless fish are groups of fish.

brook lamprey

lamprey

Fishy Fact

Fossils show that the earliest fish on Earth did not have jaws. Today, nearly all fish do. Lamprey and hagfish are now the only jawless fish.

A species is the most specific way to classify animals. There are many species within a group. Members of the same species share the same traits and are able to have babies with each other. The angelfish is one fish species.

Some types of angelfish live in the ocean, while other kinds live in freshwater.

CHAPTER 2

MEET THE FISH

Most fish are cold-blooded. That means they can't control their own body temperatures. Instead, a fish depends on its environment to **regulate** its body temperature. It warms up when the water is warm and cools down when the water grows cold.

Fishy Fact

The opah is one of the few warm-blooded fish. It lives in tropical waters around the world and can control its own body temperature.

Fish are vertebrates. That means they have backbones. Most fish have bony skeletons. Sharks, skates, and rays have skeletons made of **cartilage** instead of bone. But they are still considered vertebrates.

ray

skate

shark

All fish live in water. Fish need **oxygen** to live, but they don't breathe air. Instead, fish have gills. A gill is a special body part that takes oxygen from water, similar to how your lungs take oxygen from the air.

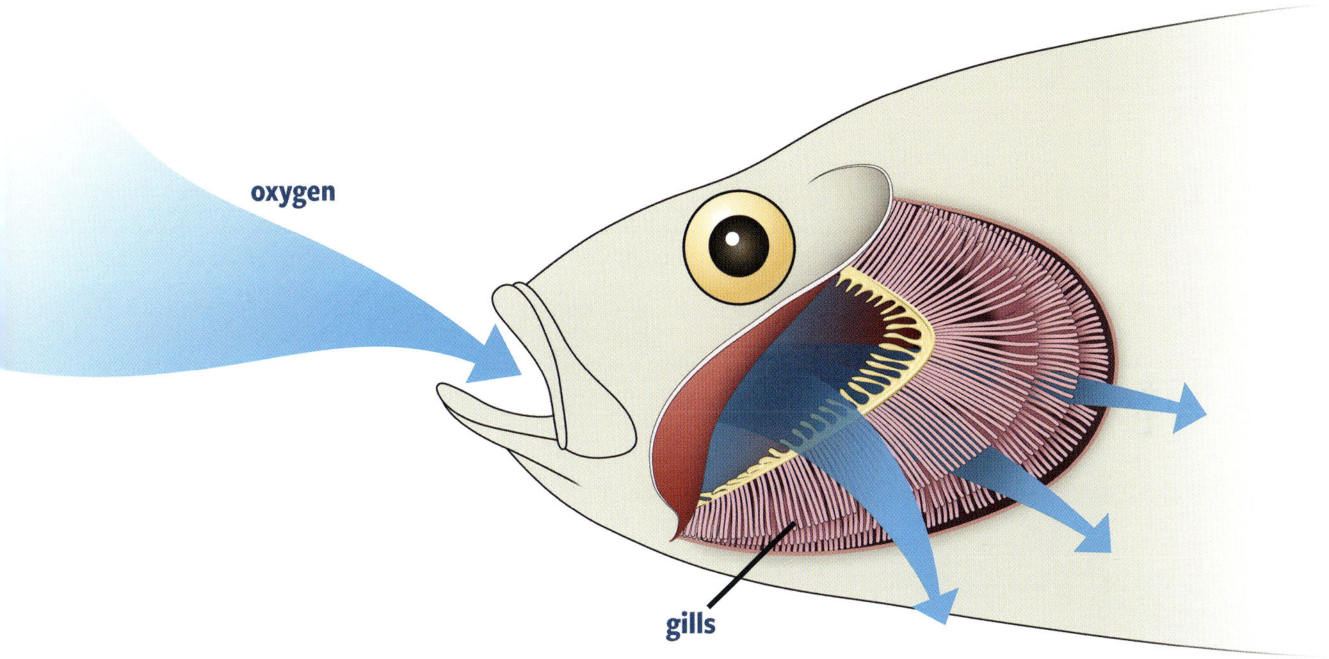

Fish have fins that help them move through water. Many fish are covered in scales. Scales are different shapes, colors, and sizes. Scales have a hard covering that helps protect the fish.

Scales grow along with a fish. Looking at a fish's scales helps scientists tell how old a fish is.

CHAPTER 3

HOME SWEET HOME

Fish are found in almost every body of water around the world. They live in tropical, **temperate**, and polar oceans. They're also found in lakes, ponds, rivers, and streams. Some fish even live under Antarctica's ice.

Fishy Fact

Antarctic toothfish have a kind of antifreeze in their blood. It allows them to live in water with freezing temperatures.

SALTWATER FISH

Ocean water contains high levels of salt. The fish that live there are called saltwater fish. Their bodies are able to get rid of any extra salt they don't need. Tuna, cod, and flounder are all saltwater fish. Colorful tropical fish found around coral reefs are also saltwater fish.

An adult yellowfin tuna can weigh more than 400 lbs (181 kg) and grow nearly 7 feet (2 m) long.

FRESHWATER FISH

Bodies of water that contain almost no salt are called freshwater. Nearly half of all fish species live in freshwater. They are often a duller color so they can blend into the bottoms of lakes or rivers and hide from predators. Bass, catfish, and trout are all freshwater fish.

Fishy Fact

Some freshwater fish are huge. The arapaima is found in Amazon rainforest rivers. It can weigh more than 400 pounds (181 kg).

BRACKISH WATER

Brackish water is found where freshwater meets saltwater, such as where a river joins the sea in an **estuary**. Some fish use estuaries as a nursery for their young. Chinook salmon swim from the ocean to lay their eggs in freshwater streams and estuaries. The young salmon return to the ocean when they are fully mature.

CHAPTER 4

GROWING UP

Most fish lay thousands of eggs. An ocean sunfish can lay up to 300 million eggs at a time! Some fish eggs float in the water. Other fish lay their eggs among plants or rocks.

fish eggs

The fish **embryo** grows inside the egg and feeds from a yolk sac attached to its body. The embryo grows into a **larva** and hatches. How long a fish takes to hatch depends on the species. Goldfish may take about a week. Rainbow trout sometimes need as long as a month.

Fishy Fact

A male betta fish blows bubbles to make a nest. When the female lays eggs, the male catches them in his mouth. Then he puts the eggs in the bubble nest and guards them until they hatch.

LARVA

A fish larva lives on its own and eats from its yolk sac. Some larvae keep the yolk sac for a few days. Others keep it for a few weeks. The larva is called a fry when the yolk sac is gone.

nile tilapia larva

FRY AND JUVENILES

A fry hunts for its own food and eats tiny insects and plants. The fry grows into a juvenile, usually in about a month. Juveniles look like adult fish, but they can't yet have young of their own.

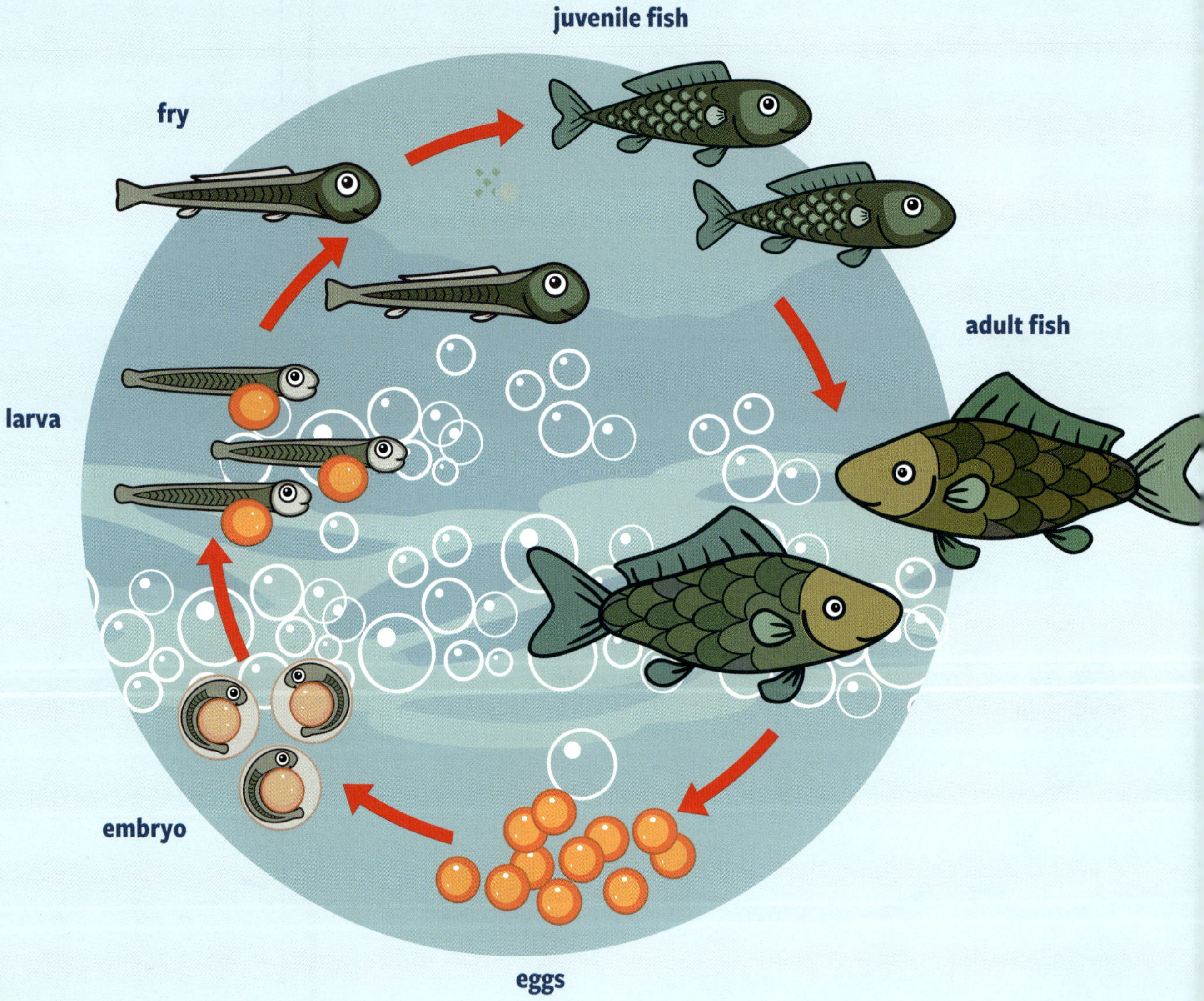

ADULTS

A fish is an adult once it's able to find a mate and have young. Some fish species reach adulthood in a few months. Others take years. A sturgeon needs at least 15 years to become an adult.

Juvenile sturgeon start out small, but they can grow up to 20 feet (6 m) long as adults.

CHAPTER 5

ON THE MOVE

Fish swim by **flexing** their bodies back and forth. Bending their bodies pushes against the water and moves the fish forward. Some fish are fast swimmers. The sailfish can swim as fast as a car driving on a highway.

Fishy Fact

Some fish also use their bodies to jump. A flying fish swims fast underwater and then launches itself out of the water. The fish fans its fins out like wings and can glide up to 650 feet (198 m) through the air.

SWIM BLADDER

Most species of bony fish have a special body part called a swim bladder. The swim bladder is in the fish's belly. The fish floats or sinks by controlling the amount of air in its swim bladder. The fish can also fill it with just enough air to float in place or glide.

Fishy Fact

Some fish also use their swim bladders to make noise. Toadfish make such loud noises with their swim bladders that people on land can hear them.

CHAPTER 6

ON THE HUNT

HERBIVORES

What fish eat depends on the species. Some kinds of fish are herbivores. They eat only plants, such as algae or seagrass. Fish that eat only plants have teeth made to shred leaves.

Fishy Fact

Yellow tang are found in tropical waters from Japan to Hawaii. They feed on algae. Yellow tang help coral reefs by eating fast-growing algae before it can overgrow the reefs.

CARNIVORES

Carnivorous fish eat meat such as **krill** and shrimp. Predatory fish are carnivores that feed on other fish. Carnivorous fish often have large mouths filled with sharp, pointed teeth that grab prey and tear off chunks. They swallow the bites whole instead of chewing.

remora

Fishy Fact

Scavenger fish feed on the bottom of a body of water. They look for dead fish, algae, or food another fish has dropped.

OMNIVORES

Many fish are omnivores. That means they eat both plants and meat. These fish have teeth for shredding plants and teeth for taking bites.

Fishy Fact

Goldfish are omnivores. They have teeth at the back of their throats. Goldfish eat plants, insects, and the eggs of other fish.

CHAPTER 7

FANTASTIC FISH

FROGFISH

Frogfish live in shallow tropical waters. They have a a rod-like body part on their heads. They wriggle the end of the rod so it looks like food to another fish. When the fish swims too close, the frogfish grabs and eats it.

The hairy frogfish's "hair" is made up of spines that help it blend in with the coral and seaweed in its habitat.

ELECTRIC EEL

Like all eels, electric eels are fish. Electric eels live in South American rivers. Special organs in their bodies produce a powerful electric charge, so they hunt by zapping their prey. A large electric eel can produce enough electricity to light 12 light bulbs!

Electric eels don't see well, so they use a weak electric signal like a radar to find their way around and to find food.

MUDSKIPPER

Mudskippers live in swamps and mud flats in Africa, Polynesia, and Australia. They breathe through gills, but they can close them to hold in water. Filling their gills with water lets mudskippers leave the water and hunt for food on land.

Male mudskippers launch themselves into the air to attract a mate, jumping more than 20 inches (60 cm) high.

LAMPREY

Lamprey are jawless fish found in rivers around the world. Some types of lamprey live in the sea. They eat by sucking the blood of other fish. Lampreys have been on Earth for more than 340 million years.

Brook lamprey only feed during the larval stage and stop eating once they become adults.

More than 18,000 different species of fish live in freshwater. Scientists think there may be at least 3.5 trillion fish in the seas. Not all species of fish have been discovered. How many kinds of fish can you name?

puffer fish

GLOSSARY

cartilage (KAR-tuh-luhj): A strong but flexible material that connects and supports parts of the body

classification (klas-uh-fuh-KAY-shun): The arrangement of things into groups according to a system

embryo (EM-bree-oh): A living creature in the earliest stage of development, before it is hatched or born

estuary (ES-choo-eh-ree): The wide part of a river where it flows into the ocean

flexing (FLEK-sing): Bending

krill (KRIL): Very small ocean creatures

larva (LAR-vuh): A stage in the development in some animals after birth or hatching and before metamorphosis

oxygen (AHK-suh-jn): A gas, present in air and water, that all living things need to survive

regulate (REG-yuh-late): To adjust the amount of something, such as heat

temperate (TEM-pr-uht): Neither very hot nor very cold

traits (TRAYTS): Qualities or characteristics that make one thing different from another

INDEX

COMPREHENSION QUESTIONS

1. Fish are cold-blooded. That means they ____.

a. can never warm up

b. never overheat

c. can't control their own body temperature

2. What do fish gills pull from the water?

a. Salt

b. Nutrients

c. Oxygen

3. What is a fish called when its yolk sac is gone?

a. A fry

b. An embryo

c. A juvenile

4. True or False: A swim bladder helps a fish float or sink.

5. True or False: Fish eat plants or meat, but never both.

Answers: 1. C, 2. C, 3. A, 4. True, 5. False

ABOUT THE AUTHOR

Tracy Vonder Brink loves to learn about science and nature. She has written many nonfiction books for kids and is a contributing editor for three children's science magazines. Tracy lives in Cincinnati, Ohio with her husband, two daughters, and two rescue dogs.

Written by: Tracy Vonder Brink
Designed by: Kathy Walsh
Series Development: James Earley
Proofreader: Janine Deschenes
Educational Consultant: Marie Lemke M.Ed.

Photographs: Shutterstock; Cover: Eugenia Petrovskaya, Joanne Weston, nicolasvoisin44, Andrej Jakubik, Joan Carles Juarez, panpilai paipa; p 2-31 Eugenia Petrovskaya; p 4 Arthur Balitskii; p 5 COULANGES, Gena Melendrez; p 6 Natalia Sidorova; p 7 @Wiki; p 8 JENG BO YUAN, Ramon Carretero, Natia Tsuleiskiri; p 9 Aldona Griskeviciene, koliw; p 10 Arcady; p 11 Steven Benton; p 12 Tom Wang; p 13 SergioRocha; p 14 Kevin Cass; p 15 Kristina Vackova, scubaluna; p 16 Produktfoto Putten; p 17 O partime photo; p 18 Kazakova Maryia; p 19 Vladimir Wrangel, Konstantin Baidin; p 20 Daniel Huebner; p21 Josephine Julian; p 22 Volodymyr Burdiak; p 23 Gary Westphalen; p 24 Mirek Kijewski; p 25 Jack FotoVerse; p 26 tristan tan; p 27 Anake Seenadee; p 28 jack perks; p 29 Moize nicolas

Crabtree Publishing

crabtreebooks.com 800-387-7650

Printed in the U.S.A./012023/CG20220815

Published in Canada
Crabtree Publishing
616 Welland Ave.
St. Catharines, Ontario
L2M 5V6

Published in the United States Crabtree Publishing
347 Fifth Ave
Suite 1402-145
New York, NY 10016

Library and Archives Canada Cataloguing in Publication
Available at Library and Archives Canada

Library of Congress Cataloging-in-Publication Data
Available at the Library of Congress

Hardcover: 978-1-0398-0025-0
Paperback: 978-1-0398-0084-7
Ebook (pdf): 978-1-0398-0203-2
Epub: 978-1-0398-0143-1